A Perfect Universe

By Thad Roberts

A Perfect Universe

By Thad Roberts

Maitreya 8808 Edition

First Print Date 03-14-2020

Published in the United States by Thad Roberts

ISBN 978-0-578-66088-2

Cover and figures by David Heggli, Thad Roberts and Jeff Chapple

Other books by Thad:
- Einstein's Intuition:
    Visualizing Nature in Eleven Dimensions
- Moon Rock: Mare Crisium
- Passages

To Einstein, Euler, and Angela.
And the combination

Chapter 1
i^introduction

*T*his book is about the origins of the nouns of reality—the things of existence, and the simplest way to get them. And it is about the verb qualities they imbue—what kind of derivatives they come with. It features the discovery of a fundamental union between physics and pure math—connecting the Planck constants to Euler's number (e). Then, before revealing the elusive pattern of the masses and unfolding into a theory of everything (toe), it offers a golden solution to an age-old mystery—a riddle that has plagued physics for over a hundred years, by deriving the value of the dimensionless fine-structure constant, from pure geometry.

For those unaware of what kind of claim is being made here, I leave a quote by the highly regarded physicist, Richard Feynman, from 1985.

"There is a most profound and beautiful question associated with the observed coupling constant e – the amplitude for a real electron to emit or absorb a real photon. It is a simple number that has been experimentally determined to be close to 0.08542455. ... It has been a mystery ever since it was discovered more than fifty years ago, and all good theoretical physicists put this number up on their wall and worry about it. Immediately you would like to know where this number for a coupling comes from: is it related to pi or perhaps to the base of natural logarithms? Nobody knows. It's one of the greatest damn

mysteries of physics: a magic number that comes to us with no understanding by man."

That statement no longer stands.

By the end of this book you will have *that* understanding. By the end of this book, you will know how to *create a universe* that gives you *that* number for free. And in that universe you will find that photons inexplicably arise, with exactly the same properties we find in our universe. Matter particles form, with exactly the same values for spin, charge, and mass as the particles in our universe. The entire set of fundamental constants of Nature pop right out—precisely matched to the ones we find in our universe. And the actions of quantum mechanics and general relativity are automatically geometrically programmed in.

Chapter 2
the simplest ontology

To end up with the most conceptually elegant and ontologically parsimonious blueprints for a universe, we need to keep the number of "things" we *call into existence* to a minimum—the minimum number needed to generate the qualities we see in our universe.

If we start off with the wrong ontology, we will have to admit more things into the universe, to make up for the mismatch between what should occur, and what actually occurs. The further we are off from an exact solution, the more *things* we will have to admit as *fundamental in the universe* to make sense of it.

Which means, we can start anywhere really—start by admitting any one thing into existence, anything you like. Just remember, every one of the Universe's character traits that your admission does not generate and explain, will require you to add at least one more number to your set of ontological furniture pieces, cluttering up a room that was designed for yoga.

If we start with the right ontology, if we perfectly identify the true foundation of reality, then we won't have to admit any other things into the universe to make sense of it. Everything will simply be naturally generated from that foundation. But how do we find that right characterization? How do we zero in on the universe's true foundational form?

Here I have a uniquely successful suggestion.

From my point of view, *being a scientist*, means having a passion for reducing your ontological number—for attempting to imagine past the things that limit the conceptual reach of your thoughts—in quest for the hidden deeper truths of *reality*. It is that *delight*—that engagement with reality—the quest to uncover more of its secrets—that connects all true scientists. The *art of science*, is the art of picking a better and better ontology, painting a story that is more of the truth over time, one that explains more with less, and offers deeper levels of explanation.

So here's my suggestion. If ontology is most beautiful when it is at a minimum, then I say let's aim to comprehend the *most beautiful universe*, whether or not that universe is ours. Let's chase the *perfect universe*, and bask in *its* ontological beauty.

That is, instead of giving up and jumping to a more complex ontology—as soon as we can't account for something, let's just stick to the plan and explore only the most elegant and parsimonious solution. Let's aim *only* for the best of all possible ontologies, and examine *its* depths. Let's see what a successful first-order ontology, really has to offer.

To get a feel for how close the modern science story gets, how close it has come to its target of achieving the true story of reality, note its current ontological ranking is about 60. So far, the scientific method has impressively whittled the number of core concepts blocking our intellectual horizon down to about 60. But 60 is still far off from 1.

Why is it 60? Because, on top of admitting one substrate—spacetime, the modern physics story gives primary ontological status to a zoo of fundamental mass particles, photons, four forces, dark matter, dark energy, dozens of parameters determined by experiment, a wave function, the Big Bang event, and depending on your philosophical leanings strings, holograms, and a few more. Which gives us a grand total of about 60.

We can do much better.

I want to present to you a different model than the one you were schooled to carry around in your brain; one that *I think you will agree* possesses ontological and parsimonious superiority—a solution from an *order one ontology*.

To build a universe, one starts by admitting *something* into existence. Let's admit a substrate, a simple quantum fluid—a superfluid.

So far our ontology number is a one—Great! But we haven't accounted for anything in the universe at all, unless you want to let that fluid's averaged-over description represent the medium of spacetime. But even then, how do we get all the rest? What ingredients do we add next to get all of the other necessary forms and properties of the universe?

Hold on. That might be how the hunt for a theory of everything usually goes, but we are not doing that this time, remember? We are *not* going to just ad hoc throw more things into the universe. This time, we are sticking to a first-order ontology. And I'll tell you how. That *one thing* we called into existence, has *two* properties. It behaves like a non-viscous fluid and possesses broken symmetries.

At first, that might sound like we have only two kinds of things to work with, but one of those things is a break in symmetry. It's a *mirror*. Do you know how many things you can get, if you start off with just two things, but one of those things is a "mirror"?

Oh, that would be a beautiful solution.

Why did I admit a *superfluid* as the first noun of reality? Why did I take that guess? Well, there's a lot of reasons really, many of them bending an ear to the pilot-wave interpretation of quantum mechanics. But one very important reason is that superfluids possess *perfectly conserved internal relations*; they have zero viscosity; and are rich in symmetry; and low in broken symmetry—possessing just one, at first, which happens to get everything else going—in a cascading fractal of broken symmetry that calls much more into form.

So admitting a perfect fluid sets us up with a substrate that is imbued with a tendency for authoring emergence—and that's exactly the kind of property we need *to get a Universe going*.

Another reason to start off with a superfluid substrate, is that this idea has been poked at before (see Schrödinger's work, or de Broglie's Double solution theory, or my earlier work on quantum space theory) with promising results. When we admit a superfluid substrate, right off the bat we get a surprise. When we derive the wave equation for that simple quantum fluid, that perfect fluid, the famous Schrödinger equation—which, according to quantum mechanics, determines how the state of the universe evolves—literally pops right out.

We will do the math later (in Chapter 10). But for now, I want to impress upon you how remarkable that fact is. We are at stage one—our very first assumption. We just said, let's assume that all we have is a simple quantum fluid—and the Schrödinger equation was thrown into our tool basket for free. And as far as universe-building tools go—that's a fantastically useful tool to have! One might take that *coincidence* as a sign that we are on the right track.

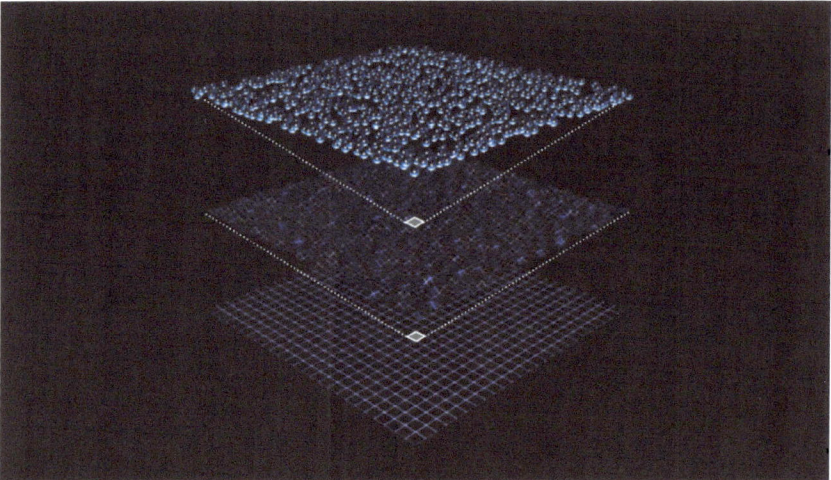

Figure 1: Scale symmetry/asymmetry
All fluid scales are similar, except those that resolve its parts.

Try to pretend that you don't know we've got the Schrödinger equation to work with. All you know is that we've admitted a perfect quantum fluid into existence, one with zero viscosity.

The structure of that fluid will be endowed with waves, fluctuations in the arrangements of its constituents—waves that show up mostly near the scales that resolve its parts. As we zoom out, the amount of *undulation* that characterizes the arena goes down. On large scales, all those little waves just average out and cancel—giving the look and feel of a medium that is waveless, empty, and boring; just a vacuous stage.

A fluid's extreme *scale-symmetry*—its ability to look the same on all scales—is broken only on the scales that resolve its parts, its microscopic scales. That break in scale-symmetry—the fact that the fluid is made up of quantized parts, is infinitely significant, because it permanently instills an imbalance into reality.

On the smallest scales there are waves, lots of waves dancing about. On large scales the medium is quiet and boring. And somewhere in-between, there's a gradient.

But so what? Why would that matter? How is this broken symmetry going to accomplish anything? Well, as it turns out, when this broken symmetry meets zero viscosity, chaos is sieved into order—in a very specific way.

At first glance you might assume that the perturbations occurring on these smallest scales will always be just random fluctuations, always scattered back to randomness. But that is *not* the case. Superfluidity encodes a tendency to transform *some* random inputs into ordered outputs.

In a superfluid, fluctuations can form harmonic relationships with other fluctuations. Stabilizing that *relationship* gives *permanence* to new form and new properties. When the right fluctuations find themselves in the right relationship, something new is born, something new is called into existence.

As we go down in scale towards more fluctuations, more contrast of waves, and a greater abundance of waves; as we go down in scale, it becomes increasingly likely that special perturbations will eventually form—waves that are capable of forming relationships with other special waves. Waves *made special* by how their mutual attempt to cancel each other, gives birth to a harmony—the kind of harmony only a superfluid can sing.

In the next chapter, we will discover how randomly induced waves set up harmonic resonances in a superfluid—that just happen to be indistinguishable from photons.

Chapter 3

simple waves—photons

"All these fifty years of pondering have not brought me any closer to answering the question, what are light quanta?"
Albert Einstein

Our experiences have given us a rich intuition for how fluids behave. When we dance our hand out the car window, and play with how the drag changes with the movements of our fingers—pulling our arm up and down with a small sway of the wrist—we develop a feel for how fluids push back, how they drag. Skip a rock on a pond and circles ripple outward from each bounce, dissipating energy, because "waves fade away".

These *intuitions*, leave us out of practice of imagining how waves would behave in a fluid that is non-viscous. As a consequence, the kinds of waves that *the absence of dampening endows a fluid with* are off our radar. We just haven't yet imagined their form. Which is why photons, and electrons, etc. seem baffling.

It's time to bridge that gap.

Let's look a little closer at our fluid. Let's specifically identify all the internal edges, or *boundaries*, that are prescribed by quantization.

Quantization prescribes more than one scale boundary. A medium made up of individual interactive spheres naturally divides into 4 domains of expectation or "action", separated by scale boundaries.

Figure 2a: Quantum scale boundary.

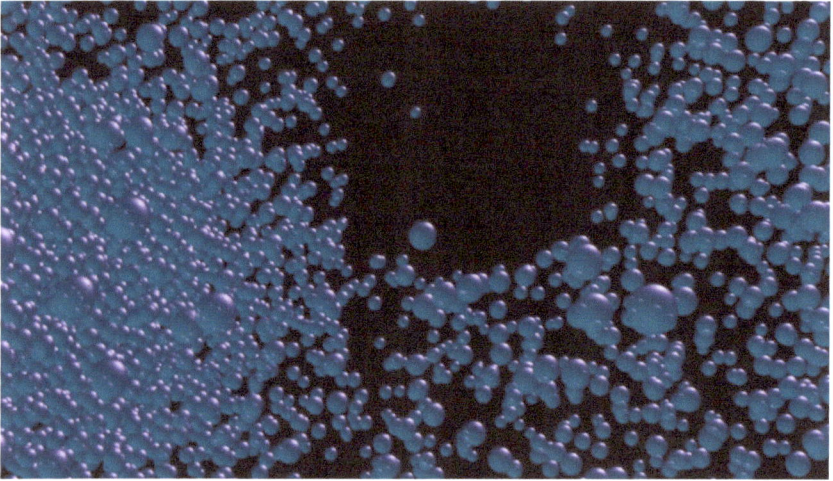

Figure 2b: Second scale boundary.

     The minimum scale boundary is set by the radius of the quanta that compose the medium—the Planck length (l_P) (Figure 2a). Going smaller than this means that there is no longer any tractable way to talk about *the fluid*. As we zoom out from this minimal scale, two more boundaries sort the internal binary structure of quantization into the smooth stage of reality we see on large scales.

The first of these boundaries is the scale in which a fluctuation in the fluid's arrangements can first be distinguished from noise (Figure 2b). This scale boundary is represented by the Planck charge (q_P). The next is the last scale that resolves the fluid as being made up of discrete parts. This scale boundary is the zero boundary of discretization—represented by the Planck mass (m_P).

Keeping these boundaries in mind, let's imagine that on a scale beyond noise scatter, a collaboration of fluctuations just happens to form a vacancy in the fluid—a region with lower density and pressure than the background. What do you expect to happen? Well, the surrounding fluid should quickly rush to fill it back in, erasing it back into noise. True. That's what will happen.

But what happens when events collaborate to make a region *lower* in density, and at the same time make another region *higher* in density? What happens if the centers of these two regions are formed in close proximity to each other, relatively at rest?

In a non-viscous fluid, when a high-density point meets up with an out of sync, out of position, out of phase low-density point, it kick-starts a harmonic resonance. Once the pair is in appropriate relation, the surrounding fluid unanimously acts to push these high and low pressure spots towards each other, to equalize things out.

The low-density spot is pushed towards the high-density spot, and the high-density spot is pushed towards the low-density spot. And as the background squeezes these distortions together, they pick up speed—a little bit of fluid picks up momentum.

When the two disturbances reach their mutual center, perfectly overlapped, their combined energy is momentarily hidden, but not deleted. In that instant, the total arrangement of the fluid's parts may look like background noise, but the arrangement of *velocities* in that system, the amount of momentum/energy collected in those parts represents a wave still in motion. That is, even though these high and low pressure bubbles are propelled towards each other, they don't stop when they reach each other. They just sail right on through, like you would expect when fluid in motion meets a *vacancy* in opposite motion.

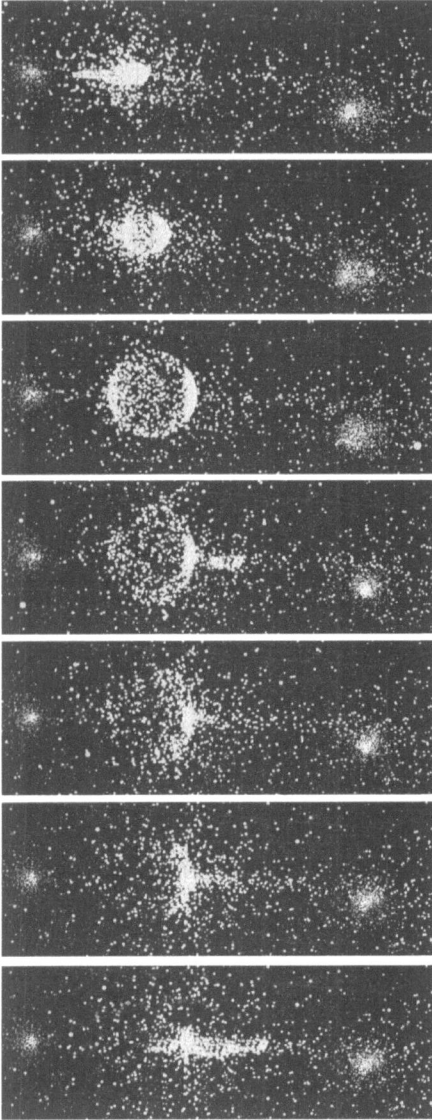

Figure 3: Simple wave in a quantized compressible non-viscous fluid.
A snapshot sequence of a simple wave.

After the moment of perfect overlap, the forward momentum of that little fluid bullet pushes on the fluid in front of it, drawing a low-pressure gap *back into existence* behind it. As it does this, it spreads its momentum forward until it becomes indistinguishable from its forward surroundings. As that indistinguishabilty comes to a head, the fluid from behind collapses forward into the gap, shooting another jet forward.

And there they go, off to the races, bouncing forth and forth, folding through each other in perfect resonance. I say "off to the races", because the center point of this resonance is carried off at the speed of sound in the medium. This little harmonic bullet, a wave that turns itself inside and out over and over, darts off one direction at full speed.

The harmony that is set up in this pairing of fluctuations creates something that *stabilizes*, granting a *relationship* permanence, calling into existence a new presence in the fluid, a new here-to-stay *form* with new properties.

This kind of wave is available only to fluids with zero viscosity. The internal drag of a viscous medium would quickly dampen this kind of resonance out.

The only reason these kinds of waves are unfamiliar to us, is that we don't have practical experiences with non-viscous fluids, so we haven't spent much time *imagining* the waves that occur in them. They are not difficult to imagine. They are *easy* to imagine. And they come with a set of properties that perfectly match the photon—something in desperate need of an accurate analogy.

In what sense do they behave as photons? Wholly and exactly. To start with, these simple waves are harmonic bullets that corpuscularly transport energy through the medium, over great distances, without loss of concentration in the energy of the wave. And they are always moving forward at the speed of sound in the fluid—the speed of light (c). They are literally incapable of just bouncing back and forth through each other around a stationary point in the fluid. Because while one half of the distortion represents *a bit of fluid* moving in the forward direction, the other represents *a vacancy* moving the other way.

A vacancy moving in the backward direction is equivalent to fluid moving in the forward direction. Which means that both of these equal but opposite parts work together to push a vibrating bubble forward.

Spend some time imagining these waves. Notice that, while the vacancy is being pulled open from one side it is also being squeezed shut from the other, as the fluid collapses behind it. While the fluid in front slows to a halt, the fluid in the back is accelerating forward. This process maintains a resonance and a symmetry. The more time you spend imagining that symmetry exchange, the more you end up intrinsically understanding photons.

By admitting a superfluid as the substrate of reality we get corpuscular harmonic waves for free; waves that transport quantized amounts of energy at the fluid's speed of sound—little packets of energy that can be sent as messengers through the fluid. These waves are composed of two undulations that oscillate in tandem, reaching their respective amplitudes at the same time, and then shrinking together until they cancel each other out momentarily. These waves never dampen, never dilute—and they come with a fixed resonance frequency.

Furthermore, the pulsing and twisting these waves impart on their surroundings as they move through it are identical to the effects we have labelled "electric" and "magnetic" fields.

To visualize what those fields are, just think of the impact a corpuscular resonant bullet has on the fluid it passes through. Due to its resonant nature, representing a *big* distortion, then a *null* distortion, then a *big* distortion, and so on, the photon is constantly adjusting the arrangements of the fluid around it. As it pulses, the surrounding fluid is pushed away from the corpuscular bit of energy, and then pulled towards it. Over and over, dragged in a cycle, with a frequency determined by the frequency of the resonant bullet.

These waves harmonically pattern the fluid around the photon into segmented domains of push and pull. And although the push and pull gets weaker and weaker at greater distances from the source, the spacing between the segmented domains remains fixed—determined by the period of the back and forth cycle.

This back and forth pulsing of the surrounding fluid is what we have historically called an "electric field".

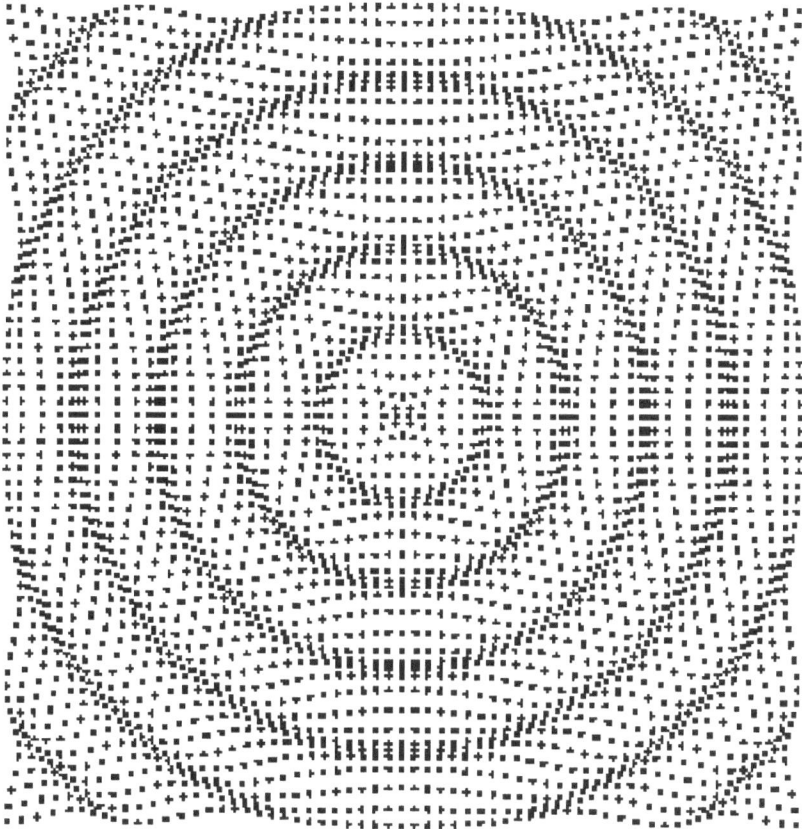

Figure 4: The electric field.
Every point is dragged back and forth radially (towards and away from the oscillating point source). This motion is the electric effect.

What about the magnetic field? Well, instead of being a measure of the push and pull *towards* or *away from* the source, magnetism is a manifestation of the *other kind of fluid distortion* that is being created as the photon goes by—twists.

A stationary electric charge patterns the fluid around it into fixed segmented boundaries. But a moving electric charge causes those boundaries to change their orientation (to twist) as the source goes by. This is why a moving electric field induces a magnetic field.

Map out the degree of twisting at each point in the fluid and now you've literally drawn the "magnetic field".

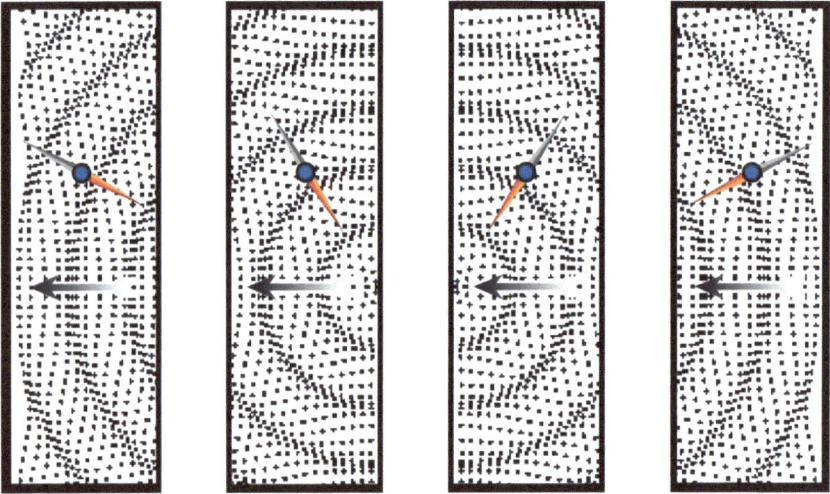

Figure 5: The magnetic field.
As the source moves, the direction each point is dragged changes. This twisting is the magnetic effect.

A photon's magnetic field is much weaker than its electric field—because the twist these harmonic bullets impart on their surroundings is far less pronounced as a distortion than the back and forth "electric" ripples that drag them.

A "point source" undergoing harmonic resonation partitions the fluid around it into standing waves. In order for this pulsing to cause a twist in the fluid, the source of that pulsing needs to be changing its angle along the horizon with respect to the observer.

There are no magnetic monopoles because in a fluid there's just nothing analogous to a "point source of twist". The very idea of twisting isn't point source available. Points have perfect spherical symmetry. So they don't select out a special direction. Twisting has an axis of symmetry, a clearly singled-out special direction.

If you've never had electromagnetism explained in terms of fluid dynamics, then I recommend you watch this fantastic 3Blue1Brown video (Divergence and Curl: The language of Maxwell's equations). Fluid dynamics isn't just "a way" one *can* talk about electromagnetism. It's *the* way to talk about it. Electromagnetism is fundamentally written in the language of fluid dynamics.

Have you ever wondered why a shorter wavelength photon represents something with more energy than a longer wavelength? Have you ever wondered why photons of smaller wavelengths have more energy? Instead of the other way around?

Well, the wavelength we've been talking about isn't the size of the distortion the quantum bullet makes as it stretches through itself cyclically. That's not the electro-magnetic field. The electromagnetic field is what persists in the surrounding fluid because of that bullet's cyclic motion. It is the pattern of segmentation that the bullet divides the fluid around it into.

The greater the energy of that central photon—the harder that jet slams against its surroundings—the more partitions the surroundings get broken into. That's why the electromagnetic wavelength associated with a photon of greater energy is smaller—greater tugs on the surroundings cause greater distortions of the surroundings, which means tearing it into more divisions of push and tug.

What were the odds that by admitting just one thing into existence we'd end up getting photons for free? I mean, we didn't just get *a mechanism for the emergence of order* (something difficult to match in value), we specifically got *photons* popping out as the very first emergent forms. And we didn't just get one or two properties of photons, we got them all at once, all in one package—explaining where they come from, what they are; exposing them as a simple, easy to conceptualize resonance sustained in a superfluid—an easy to imagine wave that accounts for every single photon property.

What were the chances that from our first ontological admission we'd end up getting a natural explanation for: how photons are sieved into reality from the edge of randomness below, why photons come in quantized amounts of energy, why they have an internal frequency, how they create the effects of electric and magnetic fields, why they can be polarized, or circularly polarized, why they can pass right through each other, why they create an interference pattern when subjected to the double-slit experiment, and on, and on? What are the chances that all of that would cascade so perfectly off our crack in symmetry, if it wasn't the right crack?—If we weren't onto the right geometry?

On the other hand, if this ontology is correct, and a quantum superfluid really is at the bottom of everything, shouldn't some very specific physical conditions follow?

Chapter 4

the fingerprint & signature
of a quantum fluid

"Philosophy is written in this grand book, the universe, which stands continually open to our gaze. But the book cannot be understood unless one first learns to comprehend the language and read the characters in which it is written. It is written in the language of mathematics…"
Galileo Galilei

One of the biggest surprises of science, mentioned far too little, is that every *unit of measure* you can think of (fortnight, decibel, tesla, bushel, femtogray, acre, light year, ohm, carat, …), every single one of them, can be expressed as a measure of length, time, mass, charge, or temperature—or a combination of those. The entire tapestry of physical reality, in all its complexity, is twisted together from just five unique *kinds of measure*![1]

Each of these *fundamental kinds of measure* has been experimentally determined to have a cut off—an absolute limit beyond which that measure is no longer a "measure". Which means that all of reality is crafted from just 5 unique *limits*.

Time and length have *minimum* limits, called the Planck time and the Planck length, whose values are

---

[1] Neither a mol or a candela makes this list. A mol is not a measure of kind, it is a measure of *amount*. And a candela is a measure of luminous power per unit solid angle emitted by a point light source in a particular direction, which can further be broken down into terms of length, time and mass.

$$t_P \rightarrow 5.391247(60) \times 10^{-44}s$$

$$l_P \rightarrow 1.616255(18) \times 10^{-35}m$$

(The two numbers inside the parentheses represent the measurement error in the two proceeding digits.)

While charge, mass and temperature have associated *maximum* limits: the Planck charge, Planck mass, and Planck temperature.

$$q_P \rightarrow 1.875545956(41) \times 10^{-18}C$$

$$m_P \rightarrow 2.176435(24) \times 10^{-8}kg$$

$$T_P \rightarrow 1.416785(16) \times 10^{32}K$$

These numbers capture the differences in scale between Nature's limits of time, distance, charge, mass, and temperature, and our references for those measures—the second, meter, Coulomb, kilogram, and Kelvin. But what are they? Why do they exist? Is there any connection between them?

If the substrate of reality is a quantum fluid, then the 5 Planck constants represent the 5 scale boundaries evoked by quantization. Which means that—together with π and ж (which encode how those boundaries stably intermix)—these numbers form the complete *base* code of reality. (See Chapter 7 for an explanation of ж.)

All of the limits and characteristics of physics are derivatives of these foundational 7 numbers. Including the constants of Nature that populate the universe with the actions of physics (more on this in the next chapter), the masses of the fundamental particles of matter, and the most famous constants of math.

To expose the beauty of this structure, note that if the substrate of reality is a quantum fluid, and if the Planck constants exhaustively encode its internal scale boundaries, then a break in scale-symmetry should directly and obviously connect the Planck constants. In other words, if the Planck constants represent the scale boundaries evoked by quantization, then they should be simply connected via a break in scale-symmetry.

Are they? Does the stamp of quantization connect the Planck constants? Here's how we check.

A break in scale-symmetry leaves a very specific and easy to recognize fingerprint. A pure break in scale-symmetry precision-splits one thing into two things, such that the two resulting things are identical in every way after the split, except in scale. If the consequences are numbers, as they are in this case, then this precision-split leaves us with two numbers that are digitally symmetric, yet distinctly different in magnitude. For example,

$$a = 4.54345754 \times 10^7$$
$$b = 4.54345754 \times 10^3$$

To find the difference in scale between any two numbers we take the log of their ratio.

$$log\left(\frac{a}{b}\right) = n$$

The output n is a whole number (positive or negative) only when the numerator and denominator share perfect digit symmetry—when they have the same digits. This means that when this equation spits out a whole number, we know the two numbers inside it are related by a break in scale-symmetry. And the magnitude of the break in scale is equal to the difference in power between them.

$$log_{10}\left(\frac{4.54345754 \times 10^7}{4.54345754 \times 10^3}\right) = 4.00000000$$

If the Planck constants are connected via a break in scale-symmetry, then they should simply connect two numbers like this—two numbers with the same digits but different whole number power exponents.

What whole number should we expect?

If we assume that a break in scale-symmetry (and only a break in scale-symmetry) is responsible for giving rise to the 7 base parameters of reality, then under symmetry minimization, n must be equal to 7.

ж, π

$$log_{10}(1 \times 10^{-7}) \quad \rightarrow \quad q_P, \; m_P, \; T_P$$

$$t_P, \; l_P$$

7 powers lost $\quad \rightarrow \quad$ *7 powers gained*

Here our ontological-least-action principle plays a guiding role in our process, pointing to the simplest, and at the same time, most symmetric solution. The one in which 7 powers (orders of magnitude) are traded (in the act of quantization), for 7 operators—powers of different kind.

You might expect such a peculiar and specific prediction to rule this ontology out—that by requiring our foundation to be that simple and that symmetric we would entirely fail to find a match with reality. That is, you might reasonably expect that the specific quantum fingerprint of a seventh order break in scale-symmetry just won't turn up anywhere in these seven parameters; that in actuality the odds of a (1×10^{-7}) being precisely stamped between the Planck constants is so low it's not even worth looking for.

Or maybe you think that the Planck constants are by definition not connected, since they relate measures that are entirely orthogonal from each other using "arbitrary" reference scales?[2]

And yet, as if by magic, the value of the Planck length multiplied by the Planck mass possesses perfect digit symmetry with the value of the Planck charge multiplied by itself (within the error bars of our best measurements).

$$l_P\, m_P = 3.517673(78) \times 10^{-43}\ m\ kg$$

$$q_P{}^2 = 3.51767263(15) \times 10^{-36}\ C$$

And the power difference between these digit symmetric numbers is exactly 7.

$$log_{10}\left(\frac{l_P\, m_P}{q_P{}^2}\right) = -7.000000\ ...$$

$$\frac{l_P m_P}{q_P{}^2} = 1 \times 10^{-7}$$

The Planck constants, the physical limits of measure that compose our reality, are indeed broadcasting a break in scale-symmetry of 7 orders. A 7 is power-scripted into the scale structure of the base code of physical reality—self-referentially embedded!

---

[2] This is a common misconception. While it is true that the meter is an arbitrary human reference scale for length, and the second is an arbitrary standard of time, and so on, the arbitrariness in each case is identical. They form a *coherent system of units* that together arbitrarily reflect the human scale. This matters because identically arbitrary numbers cancel their arbitrariness when combined.

And that's not all that connects the Planck constants.

If the foundational script of reality includes 5 limits of different measure, then the blended union of those 5 limits should intrinsically describe Nature's base of entropic action. That union should be central to every-thing, every-action, every-where. Which means, among other things, that it should be something of importance and profound elegance. Is it? Let's check.

To find the unordered union of the five Planck constants we multiply their *strings of numeric digits* together (yielding a new string of digits—the product of the union), and we sum the absolute value magnitude of each power (exponent), to get the maximum range of power in the union. Then we take the ratio of the two.

t_P	5.391245(60)	$\times 10^{-44}s$	$\rightarrow$	44
l_P	1.616255(18)	$\times 10^{-35}m$	$\rightarrow$	35
q_P	1.875545956(41)	$\times 10^{-18}C$	$\rightarrow$	18
m_P	2.176435(24)	$\times 10^{-8}kg$	$\rightarrow$	8
T_P	1.416808(33)	$\times 10^{32}K$	$\rightarrow$	32

$$\times \ \overline{\qquad\qquad\qquad} \qquad\qquad +\ \underline{\quad}$$

$$= 50.394525\ldots \qquad\qquad\qquad\qquad = 137$$

The product of the digit strings and the sum of powers are,

$$\prod \text{digits}_P \ = 50.393726\ldots$$

$$\sum |\text{powers}_P| = 137$$

And the union of those two numbers—the ratio of that power to product is equal to e (Euler's number).[3]

$$\frac{\Sigma|\text{powers}_P|}{\Pi \, \text{digits}_P} = \frac{137}{50.3937(22)} = 2.71859(12)$$

$$\cup \, (t_P, l_P, q_P, m_P, T_P) = e$$

Galileo was profoundly right. Mathematics isn't just a tool we use to uncover the hidden secrets of our physical world. The physical world and mathematics are mutual derivatives of each other. They give rise to one another. The actions of physics are beautifully orchestrated by mathematics. And at the same time, the most elegant base of action in mathematics—e, representing the natural base of rotation, exponentiation, and growth—is scripted by the union of reality's 5 *physical* limits.

$$e^{\pi i} + 1 = 0 \qquad\qquad e^{\pi} = (-1)^{-i}$$

$$e = \sum_{n=0}^{\infty} \frac{1}{n!} \qquad\qquad \int_{1}^{e} \frac{1}{t} dt = 1$$

$$e = \lim_{n \to \infty} \frac{n}{\sqrt[n]{n!}} \qquad\qquad e = \lim_{n \to \infty} \left(1 + \frac{1}{n}\right)^{n}$$

---

[3] Exact precision here requires inclusion of a second order term for mixing 2 elements of momentum between the bounds of e and 2π over the 7 scales of interaction. See Chapter 9 for the conjugate of that operation.

$$\cup \, (t_P, l_P, q_P, m_P, T_P) = e \left(\frac{(e - 2 + 2\pi)}{7}\right)^{1/2}$$

$$e = \frac{\Gamma(x+1,-1)}{\Gamma(x+1)} \frac{x!}{!x} \qquad \phi(x) = \frac{1}{\sqrt{2\pi}} e^{x^2/2}$$

What could be more elegant than the fact that Euler's formula, *commonly voted the most beautiful formula in all of math*, also stamps the vertex of reality's physical limits? What could be more delicious, than the fact that the span of physical action, throughout the entire domain of objective measure; the entire personality of existence—comes down to the special base of action where a π-rotation is equal to a reflection?

$$e^{\pi i} = -1$$

To get a feel for the richness encoded by this action base, a taste of the delicious potential enfolded by this *fact of the universe*, notice that one raised to its negative square root power, remains unchanged.

$$(1)^{-\sqrt{1}} = 1$$

While negative one raised to its negative square root power, equals the Planck union—Nature's limits of measure blended together—raised to the power of π.

$$(-1)^{-i} = e^{\pi}$$

Recall that $i = \sqrt{-1}$.

The fact that Euler's formula ($e^{\pi i} + 1 = 0$), elegantly relates the 5 most central numbers in math—*and* the 5 most foundational numbers in physics, really drives the point home that the relationship between physics and pure math runs to the very bottom, of well, everything.

The union of physical limits, rotated by πi radians (twisted half way around a circle in the complex plane), is equal to the simplest of all reflections (-1).

$$\left(\frac{\sum|\text{powers}_P|}{\prod\text{digits}_P}\right)^{\pi i} = -1$$

Chapter 5

the constants of Nature

$\mathcal{T}$he constants of Nature (the speed of light, the gravitational constant, the magnetic constant, etc.) are a set of experimentally determined parameters that have been found to define reality's fixed measurable character traits. These characteristics have been observed to always hold true, everywhere—throughout the entire domain of existence (the entire fluid we call the vacuum).

Explaining the constants of Nature is one of the first things a theory of everything would like to do. So let's do that next.

If the foundation of reality is fully characterized by 7 parameters—5 that script its boundaries of quantization $(t_P, l_P, q_P, m_P, T_P)$ and 2 that control how those boundaries intermix $(\pi, \text{ж})$, then the constants of Nature should be simply determined by the intersections of these 7 numbers. Nature's most fundamental character traits should be simple and direct references to how these boundaries interact. The values of those boundaries are:

$$T_P = 1.416785(16) \times 10^{32} \ K$$
$$m_P = 2.176435 \times 10^{-8} \ kg$$
$$q_P = 1.875545956(41) \times 10^{-18} \ C$$
$$l_P = 1.616255(18) \times 10^{-35} \ m$$
$$t_P = 5.391247(60) \times 10^{-44} \ s$$

$$\pi = 3.141592653589\ldots$$
$$\text{ж} = 0.085424543153\ldots$$

These 7 quantization parameters are responsible for the constants of Nature in the most simple and straightforward way. For example, since speed is a measure of distance divided by a measure of time, Nature's limiting speed should be the quantum limit of distance divided by the quantum limit of time—the Planck length divided by the Planck time. What is that speed? The speed of light. Exactly. And that's how simple it is for the rest of the constants of Nature too. They are all simply and naturally prescribed in reference to these same 7 parameters.

constant symbol & name value (in metric units)	origin
c (speed of light) 2.99792458×10^{8} m/s	$\dfrac{\ell_P}{t_P}$
$\hbar$ (Planck's constant) $1.054571726(47) \times 10^{-34}$ m²kg/s	$\dfrac{\ell_P^{2}\, m_P}{t_P}$
G (gravitational constant) $6.67384(80) \times 10^{-11}$ m³/kg s²	$\dfrac{\ell_P^{3}}{m_P\, t_P^{2}}$
α (fine-structure constant) $7.2973525698(24) \times 10^{-3}$	$Ж^{2}$
e (elementary charge) $1.602176565(35) \times 10^{-19}$ C	$Ж\, q_P$
k (Boltzmann constant) $1.3806488(13) \times 10^{-23}$ m²kg/s²K	$\dfrac{\ell_P^{2}\, m_P}{t_P^{2}\, T_P}$
μ_0 (magnetic constant) $1.25663706143... \times 10^{-6}$ m kg/C²	$\dfrac{4\pi\, \ell_P\, m_P}{q_P^{2}}$
ε_0 (electric constant) $8.854187817... \times 10^{-12}$ s²C²/m³kg	$\dfrac{t_P^{2}\, q_P^{2}}{4\pi\, \ell_P^{3}\, m_P}$
κ (Coulomb's constant) $8.9875517873... \times 10^{9}$ m³kg/s²C²	$\dfrac{\ell_P^{3}\, m_P}{t_P^{2}\, q_P^{2}}$

Symbol	Description	Formula
σ	(Stefan-Boltzmann constant $5.670373(21)\times10^{-8}$ kg/s³K⁴	$\dfrac{\pi^2 m_P}{60 t_P^3 T_P^4}$
R_K	(von Klitzing constant) $2.58128074434(84)\times10^4$ m²kg/sC²	$\dfrac{2\pi \ell_P^2 m_P}{Ж^2 t_P q_P^2}$
K_J	(Josephson constant) $4.83597870(11)\times10^{14}$ sC/m²kg	$\dfrac{Ж\, t_P q_P}{\pi \ell_P^2 m_P}$
Φ_0	(magnetic flux constant) $2.067833758(46)\times10^{-15}$ m²kg/sC	$\dfrac{\pi \ell_P^2 m_P}{Ж\, t_P q_P}$
Z_0	(characteristic impedance) $3.767303134617...\times10^2$ m²kg/sC²	$\dfrac{4\pi \ell_P^2 m_P}{t_P q_P^2}$
G_0	(conductance quantum) $7.7480917346(25)\times10^{-5}$ sC²/m²kg	$\dfrac{Ж^2 t_P q_P^2}{\pi \ell_P^2 m_P}$
H_C	(quantized Hall conductance) $3.87404614(17)\times10^{-5}$ C²/m²kg	$\dfrac{Ж^2 t_P q_P^2}{2\pi \ell_P^2 m_P}$
c_1	(first radiation constant) $3.74177153(17)\times10^{-16}$ m⁴/kg s³	$\dfrac{4\pi^2 \ell_P^4 m_P}{t_P^3}$
c_{1L}	(spectral radiance constant) $1.191042869(53)\times10^{-16}$ m⁴kg/s³	$\dfrac{4\pi \ell_P^4 m_P}{t_P^3}$
c_2	(second radiation constant) $1.4387770(13)\times10^{-2}$ m K	$2\pi \ell_P T_P$

When we start from a quantum fluid, sorting out why all the constants of Nature are what they are becomes astonishingly simple. Reality's diverse character traits are naturally, simply, and exactly scripted by the intersections of the limits that characterize the boundaries of quantization. The constants of Nature are simply derivatives of reality's quantum geometry.

Let's do a quick review.

So far, we have found that the 7 parameters required to characterize a quantum superfluid (coding its break in scale-symmetry), also give us all the constants of Nature—exactly—as their natural derivatives. And, just to make a show of elegance, the limits of physical measure among those parameters happen to have a union equal to e, which is in fact the base of action in our world.

From there, we discover that as we zoom out, the unique properties of this fluid inescapably lead to the emergence of self-sustaining resonances—simple waves, which simultaneously produce photons (along with their electromagnetic waves) with precision.

Okay, that's certainly a promising start. But where do we go from here?

Chapter 6
the ladder of emergence

*I*n the simplest ontology, each level of emergence becomes populated with essences that are nothing other than the stable relationships that form between the already present lower level constituent entities (including relationships between the levels). On each scale, the properties that emerge will always be the new ways that zooming out in scale builds additional cyclical stability into the existing arrangements and behaviors of the system's lower level constituents.

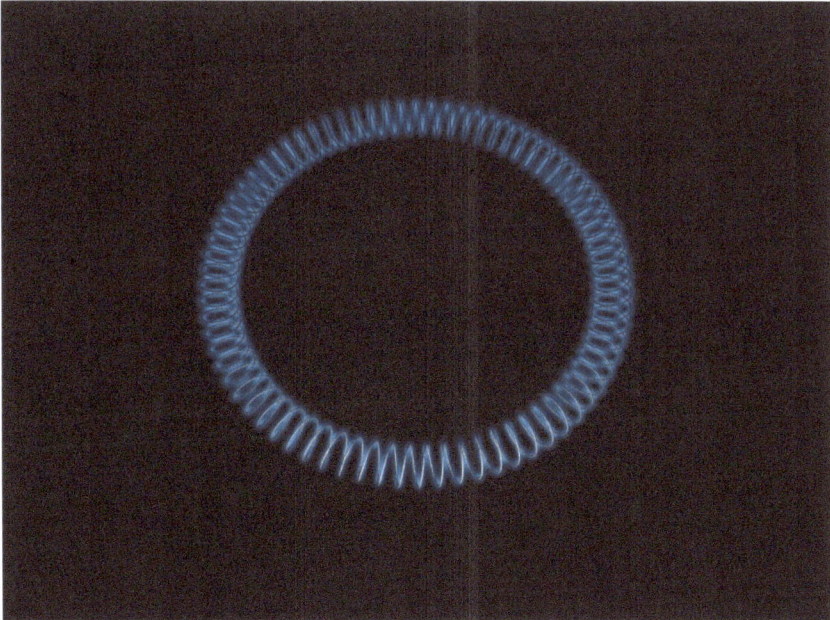

Figure 6: *A closed-loop photon path.*

Given that the nouns that pop up in each level of emergence will always be successful periodic stabilities that form between existing forms, the simplest possibility for matter particles, is that they are what happens when photons are successfully twisted around a closed loop path, forming a new periodic phenomenon.

Forming this new stability breaks another symmetry (the symmetry of having photons zip around in a way that can be described as random). On the other hand, accomplishing this action gives birth to the presence of a standing wave in the medium—a new cyclical phenomenon.

What specific *properties* are called into existence by this stabilization? What properties automatically emerge when photons manage to twist into a stable closed loop?

First off, this closed-loop cycle gives persistence to something with momentum that isn't zipping through the fluid at the speed of sound. Its internal photon is zipping around at the speed of sound, but it (the closed loop) isn't. That is a fairly *big* deal. Because the ability to carry localized momentum is literally the property we call "mass".

Second, as a photon traces around a vortex path, its electromagnetic fields twist about a pole. This twisting charges the region with a new kind of stable flow, a new property called "spin"—a signature of vorticity, which also happens to be a fundamental property of mass particles.

Third, stabilizing a twist around a torus loop creates a bounded region in which a new kind of distortion dominates—a geometric boundary in which the electro-magnetic distortion is magnified. In the smaller circle of that torus trace this distortion is then further intensified by an amount set by the ratio between the sizes of the two circles that make up the vortex's representative torus trace.

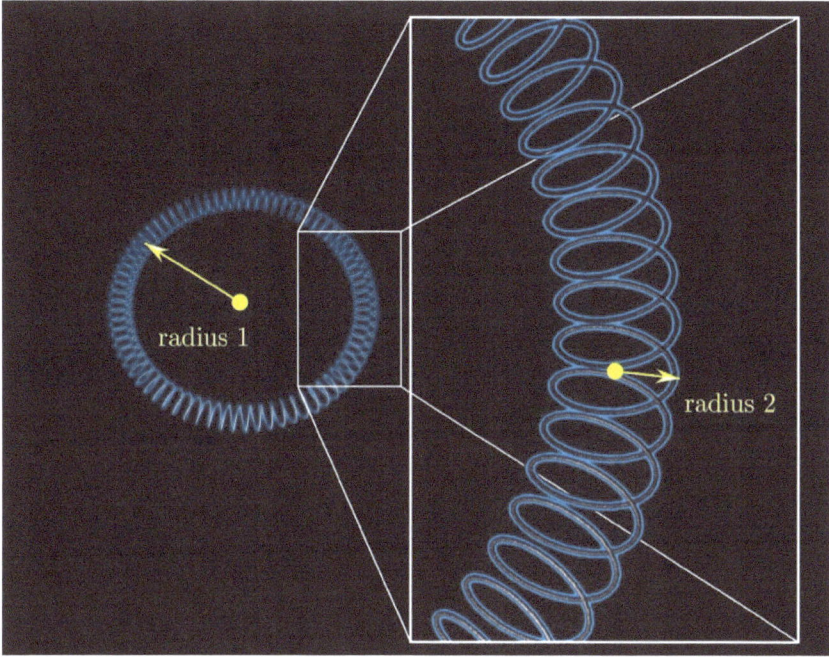

Figure 7: Closed loop.
When a photon traces a torus vortex path its momentum is confined by two different radii.

Let's characterize this vortex.

Chapter 7
ж – and the si^{i}mple vortex

"Turbulence is the most important unsolved problem of classical physics."
Richard Feynman

*I*n this chapter, we are going to use an abstract approach (using geometry to make topological conclusions) to derive a general theorem about the character of the simplest vortex a quantum non-viscous fluid can sustain. That topological conclusion will turn out to prescribe ж, the "magic number of physics"—the fine structure constant, similar to how π prescribes circles.

To start, we note that a vortex (think stabilized smoke ring) can be equally characterized as (1) the union of two orthogonal transformations—like 2 orthogonal twists, or a simultaneous twist and a phase shift, or (2) as the intersection of three orthogonal waves.

The only way these *2 coincident orthogonal geometric operations*, and *3 coincident orthogonal waves*, can simultaneously occur from just 2 broken symmetries, is for them to be in a simple fixed relationship. Encoding this condition enables us to mathematically characterize our vortex.

To that end, let us represent a general expression for the combination of two simultaneous, orthogonal, yet *simply connected* (distinguishable by only one difference) transformations. Let one of those operations be captured as a twist, e^{-x} by some argument x, while the other transformation (already being different in kind) is represented simply by the same argument x. Now let the expression for the combination of these two operations be,

$$(e^{-x})^{-x}$$

The identity between the arguments guarantees that the two contributing components are simply connected (across their break in symmetry).[4]

In its current form, this expression captures a general geometric property between two simply connected, orthogonal (a condition guaranteed by the difference in kind) simultaneous transformations; like a twist matched in argument with a change in phase.

To make this condition capture something more specific, let's encode our stable vortex condition by setting $x = -\dfrac{\pi}{2}$, matching a quarter turn twist with a quarter cycle phase change—which we will call *the Möbius condition*.

$$\left(e^{-\pi/2}\right)^{-\pi/2}$$

Since $e^{-\pi/2} = i^i$, we can also write this matching condition as,

$$\left(i^i\right)^{-\pi/2}$$

This dimensionless number represents the amount of deformation (the amount of twist and phase rotation) that an ideal vortex imparts to a non-viscous fluid that is *continuous* on all scales.[5]

---

[4] Since only one break in symmetry is allowed to distinguish the two operators, choosing to express them as different kinds of geometric operations immediately restricts us to arguments that match.

[5] We might think of this number as relating to things that are stabilized under Möbius action, or transformation. Like going

To get the number that represents the twist and phase-shift a vortex imparts to a non-viscous *quantized* fluid, we must remember to model the core of our vortex—its throat—as an absence. This gap in the fluid is a stabilized internal boundary beyond which no fluid exists, which means it is a region outside of the domain of the waves it supports. It must, therefore, be counted as an absence—subtracted from the domain of our expression.

Accounting for that central hole, we get an expression for the distortive effects that an ideal Platonic vortex dynamically charges the non-viscous quantized fluid with.

$$\left(i^i\right)^{-\pi/2} - m_P$$

Where m_P is a dimensionless scaled parameter—the contribution value of the smallest piece of the fluid (the Planck mass) compared to the scale of reference used (from the Planck scale to the kilogram scale).

This completes our first characterization.

Since we can also generally characterize our vortex as the intersection of 3 orthogonal waves, our next step is to set the contributions of those 3 waves equal to our first characterization.

$$\lambda_1 \hat{\imath} + \lambda_2 \hat{\jmath} + \lambda_3 \hat{k} = \left(i^i\right)^{-\pi/2} - m_P$$

This equation is a simple transform between two expressions. It reads, "Three perpendicular waves intersect to form a vortex in a quantized medium when their contributions sum to the amount of matched twist-phase rotation needed to stabilize that vortex."

---

around a Möbius strip.

Right now, this equation has 3 unknowns. To solve it, there needs to be a simple relationship between those unknowns. For our solution to match our starting assumptions, that relationship must be a break in symmetry and there must be one break in symmetry upsetting that break-in-symmetry-relationship. In other words, there must be a break in symmetry on top of a break in symmetry.

Let's stipulate the break in symmetry defining the relationship between the 3 contributors before we worry about breaking that symmetry.

To stay in line with the simplest ontology, and because all we have to work with so far is a break in scale-symmetry, let's say that these three contributors are connected via a break in scale-symmetry. And since we are attempting to describe the second level of emergent forms, let's assume that the participants in this relationship are all separated in scale by 2 orders. Like this.

$$ж^{-1} + ж^{1} + ж^{3}$$

Encoding an intrinsic and fundamental interaction between 3 things that differ by powers of 2.

Why did I pick this exact arrangement instead of $(ж^{0} + ж^{2} + ж^{4})$ or $(ж^{1} + ж^{3} + ж^{5})$? Because part of the expectation of simplicity is to keep the magnitude of the powers to a minimum. Higher order terms require more to stabilize. Therefore, two first order terms and a third order term is the lowest order arrangement available. This brings us to the following equation.

$$\frac{1}{ж} + ж + ж^{3} = \left(i^{i}\right)^{-\pi/2} - m_{P}$$

In its current form, this equation relates 3 amplitudes connected by a second-order break in symmetry (differing by powers of 2), to the amount of twist and phase change needed to geometrically stabilize a vortex in a quantized non-viscous fluid.

To capture the condition we are after, all we have left to do is break the symmetry on the left side once. This is accomplished by dividing the nonlinear term by 2π, making it proportionally connected to the linear terms via circular transform—in which case it makes one whole contribution of its scale every time the others go 2π times the contribution of their scales.

Having this break in symmetry transform equally across both of the other scales allows us to keep the three contributors in constant orthogonal arrangement—yielding an ideal represent-ation of a vortex in a quantized non-viscous medium, composed of 3 orthogonal components that are symmetrically and harmonically connected.

$$\frac{1}{\text{ж}} + \text{ж} + \frac{\text{ж}^3}{2\pi} = \left(i^i\right)^{-\pi/2} - m_P$$

And we have achieved our goal!

This equation encodes the energy-conserved fluid-dynamic stabilization condition for a simple vortex in a quantized non-viscous fluid. It characterizes the simplest Platonic quantum fluid form. Just as π is a geometric encoder of circles, ж is a geometric encoder of Nature's simplest vortex.[6]

The dimensionless scale parameter in this equation (m_P) makes ж a scale dependent value—a dimensionless geometric parameter whose numeric value scales as m_P goes from 1 on the Planck scale, to $2.17651(13) \times 10^{-8}$ on the kilogram scale. Which means the value of this geometric operator (ж) scales between two limits, depending on the scale it is observed on.

---

[6] There are 4 minimum and 4 maximum solutions to this equation (two Real and two Complex each)—making it something with 8 limiting solutions. See appendix ж₇ for more details.

Those limits are:

$$ж_{max} = 0.093474041000\ldots$$

$$ж_{min} = 0.085424543153\ldots$$

$ж_{max}$ correlates with the Planck scale, where $m_P = 1$, and $ж_{min}$ correlates with the kilogram scale where $m_P = 2.176435(24) \times 10^{-8}$.

Compare this to our best-measured value of the "magic number" Richard Feynman was talking about.

$$ж = 0.085424543135(14)$$

Which, today, is understood to be a scale dependent coupling constant, and has been measured to swing up to $\approx .0887$ on energy scales of the Z boson (about 90 GeV).

"You might say the "hand of God" wrote that number, and "we don't know how He pushed his pencil." We know that kind of a dance to do experimentally to measure this number very accurately, but we don't know what kind of dance to do on the computer to make this number come out, without putting it in secretly!"
Richard Feynman

I submit, for your review, that the dance that gives us this number is now made clear.

Where does ж show up in reality? What does it help us explain? Well, for starters, it fixes the electron charge, and both of its radii (the Bohr radius (a_0), and the classical electron radius (r_e)) and the difference between those radii,

$$e = \text{ж}\, q_P \qquad\qquad a_0 = \frac{l_P}{\text{ж}^2}\frac{m_P}{m_e}$$

$$r_e = \text{ж}^2\, l_P\, \frac{m_P}{m_e} \qquad\qquad \frac{r_e}{a_0} = \text{ж}^4$$

Where l_P is the Planck length, m_P is the Planck mass, q_P is the Planck charge, and m_e is the mass of the electron.

Of course, once we have the Bohr radius (a_0), we also have the helium (He^+) radius, and the lithium (Li^{2+}) radius, and so on.

$$He^+ = \frac{a_0}{2} \qquad\qquad Li^{2+} = \frac{a_0}{3}$$

ж also sets the strength of the strong nuclear force compared to the electromagnetic force.

$$\text{ж}^2 = \frac{electromagnetic\ force}{strong\ nuclear\ force}$$

In a torus vortex, everything that flows around the large circle's boundary must also flow around the small circle's boundary. The difference in scale between those circles acts as a geometric lens, magnifying its fluid effects inside the small circle.

This gives us a simple explanation for why the strong force is $\frac{1}{\text{ж}^2}$ stronger than the electromagnetic force. Within the domain of the smaller circle electromagnetic distortions are being focused by the proportion unique to the simple vortex.

The electron's coupling parameter (ж) gives it away as a rendition of an ontologically minimum platonic vortex in a quantized non-viscous medium—which, we might note, satisfactorily settles the question about why the electron has two radii in the first place.

The electron is the simplest of all stabilized fluid forms—a vortex in a non-viscous fluid hosting a single break in symmetry, served with a Möbius twist—a spinor.[7]

All of this might prompt us to ask, are the following *equations of math*, or *equations of physics*?

$$e^{\pi i} + 1 = 0 \qquad\qquad e = 1 + \frac{1}{1!} + \frac{1}{2!} + \frac{1}{3!} \cdots$$

$$e = \Gamma(3, -1) \qquad\qquad e = \cup\,(t_P, l_P, q_P, m_P, T_P)$$

$$\int_1^e \frac{1}{t}\,dt = 1 \qquad\qquad e = \lim_{n \to \infty}\left(1 + \frac{1}{n}\right)^n$$

$$e = \lim_{n \to \infty} \frac{n}{\sqrt[n]{n!}} \qquad\qquad e = \frac{\Gamma(x+1, -1)}{\Gamma(x+1)}\,\frac{x!}{!\,x}$$

$$e = \sum_{n=0}^{\infty} \frac{1}{n!} \qquad\qquad \frac{1}{\text{ж}} + \text{ж} + \frac{\text{ж}^3}{2\pi} = \left(i^i\right)^{-\pi/2} - m_P$$

See Appendix ж₇ to further explore the rich structure contained within the simple vortex equation.

---

[7] Note how lucidly accessible spinors become in this picture.

Chapter 8

π and ж are lovers

To: Archimedes, Thales, Pythagoras, Euclid, Euler, Planck and Einstein…

Subject: Update

Dear esteemed friends,

I write to you today to give you an update. The world of geometry has gained something new. I have for you a tasty morsel that I promise will add a delightful twist to the story of reality— aligning our comprehension with the other ends of its form; which turns out to change, well—everything.

After two decades of searching for a way to encode the maximum curvature of spacetime, the answer now seems so simple. I was so close for so long. I just needed to think higher dimensionally.

To represent the two limits of curvature using the simplest of geometric forms, we perform two *different* operations—using circles. To measure curvature approaching the minimum limit (zero), we use just *one* circle. Take the ratio of the circle's two lengths, its circumference to its diameter, and a measure of π proves you are in a region with zero curvature.

But, to measure the other end of curvature—the *maximum* internally sustainable amount—you need to compare two *different* circles—the radii of the two circles of a torus. And not just any torus, the simplest harmonic torus possible—the Platonically ideal torus, riding just two breaks in symmetry. The simplest of all closed *fluid* forms.

The dimensionless geometric number that encodes the geometry of that stability condition, is a scale-dependent transcendental number.

$$\frac{1}{Ж} + Ж + \frac{Ж^3}{2\pi} = \left(i^i\right)^{-\pi/2} - m_P$$

$$Ж_{max} = 0.093474041000 \ldots$$

$$Ж_{min} = 0.085424543153 \ldots$$

This flexible number exists in opposition to π; kept in eternal relationship with it, as two ends of one physical form. The whole of reality—its every intersection—fits within the span of their transcendental connection.

To find this magic number (which turns out to be the coupling constant associated with the electron), one only needs to ask what would happen if we assumed that the composite orthogonal waves in a stable torus vortex were related by just two breaks in symmetry—harmonically twisting both circles into existence. In a quantized fluid, these conditions lead to the equation above.

π and Ж are lovers that dance through the i of reality; forever connected by reflections and twists, giving rise to thoughts and plots that enable the entire story of existence. The boundaries of quantization are responsible for all the emergent forms of reality.

This was a missing fundamental equation about the geometry of the universe.

Anyway, I thought you'd like to know. π isn't alone anymore. He never was. I just found his higher dimensional lover, and she's gorgeous, and just as endlessly endowed with numbers as he is, but a bit more flexible. You should see how over the moon he is with her. Together they are capable of e .

Sincerely,

Thad Roberts

P.S. Euler, and Planck, there's been a sort of marriage that the two of you should know about. The five natural limits of physical measure (the Planck constants) have a power to product union equal (to first approximation) to e.

$$\frac{\sum |\text{powers}_P|}{\prod \text{digits}_P} \approx e$$

$$\cup (t_P, l_P, q_P, m_P, T_P) \approx e$$

Which means that the most famous equation in all of math

$$e^{\pi i} + 1 = 0$$

relates the five most basic symbols in mathematics *and* Nature's five physical limits of measure. The link between physics and pure mathematics runs to the very limits of reality, coloring them both more beautiful.

P.P.S. I owe you a dinner Euler. You have been of extreme help. If it weren't for you, I surely would not have figured this out. And Einstein, if it weren't for you, I might not have tried.

P.P.P.S. Enjoy finding how the elusive pattern of the masses pours out of this. You've got the whole universe coming. And you're not going to believe how beautifully fractal it is.

Chapter 9

mass and charge

"One might as well speculate on the origins of matter."
Charles Darwin

$\mathcal{I}$n Chapter 4 we learned that e designates the unordered union of the 5 Planck constants and therefore applies to anything tending to disorder. But what about things that maintain their sense of order? Like vortices? How do we characterize a sustained ordered mixing of those boundaries?

To answer that question, the first thing we do is recognize that quantization constrains vortex stability to being that which balances Nature's 3 maximum boundaries against its 2 minimum boundaries. To discover how this stable ordering condition geometrically restrains the world it twists together, we divide the union of Nature's 3 maximum boundaries (q_P, m_P, T_P)—by the union of its 2 minimum boundaries (t_P, l_P).

$$\frac{\cup\,(q_P, m_P, T_P)}{\cup\,(t_P, l_P)} = \frac{7}{2^2\,(\alpha_F)^{1/2}}$$

Where α_F is the Feigenbaum Alpha constant.

What we find, is that Nature's ordered union balances its 3 maximum boundaries against its 2 minimum boundaries in a way that divides up the fluid's 7 asymmetric scales with a doubling-squaring action $(2)^2$, and a square root period doubling bifurcation $(\alpha_F)^{1/2}$. These conditions, combined with our new understanding of vorticity, turn out to be exactly the information we needed to solve the infamous "riddle of the masses".

By *generation*, mass maintains the basic vortex pattern (relating 3 things that differ by *powers* of 2), by grouping 3 groups of 3 particles together (in terms of scale) through ascending *multiples* of 2, and another 3 groups of 3 particles via a *power tower nesting* of 2.

$$3^2(m_H - m_Z) = m_W \ (2^{-1}) \ (\mu + 3 + \pi)$$

$$3 \ (m_N - m_P) = m_e \ (2^0) \ (\mu + 3 + \pi)$$

$$3 \ (m_d - m_u) = m_e \ (2^1) \ (\mu + 3 + \pi)$$

$$3 \ (m_c - 2m_s) = m_\mu \ (2^2) \ (\mu + 3 + \pi)$$

$$3 \ (m_t - m_\tau) = m_b \ \left(2^{2^2}\right) (\mu + 3 + \pi)$$

$$3 \ (m_{\nu_\tau} - m_{\nu_\mu}) = m_{\nu_e} \left(2^{2^{2^2}}\right)(\mu + 3 + \pi)$$

Where m_H, m_Z, m_W, m_N, m_P, m_e, m_d, $m_{u'}$, m_c, m_s, m_μ, m_t, m_τ, m_b, m_{ν_τ}, m_{ν_μ}, and m_{ν_e} are respectively the masses of the: Higgs boson, Z boson, W boson, neutron, proton, electron, down quark, up quark, charm quark, strange quark, muon, truth quark, tau quark, beauty quark, tau neutrino, mu neutrino, and the electron neutrino. And μ is the nontrivial zero of the logarithmic integral—not to be confused with the muon charge (μ) below.[8]

---

[8] ($\mu + 3 + \pi$) characterizes the interplay of three components of momentum bounded between an inner boundary (represented by μ,

$m_H = 125.09$ GeV/c²
125.18(16)

$m_Z = 91.1871$ GeV/c²
91.1876(21)

$m_W = 80.377$ GeV/c²
80.379(12)

$m_N = 939.56542086$ MeV/c²
939.56542047(54)

$m_P = 938.27208814$ MeV/c²
938.27208816(29)

$m_e = 0.51099895001$ MeV/c²
0.51099895000(15)

$m_d = 4.7$ MeV/c²
4.7(04)

$m_u = 2.1$ MeV/c²
2.2(05)

$m_c = 1.264$ GeV/c²
1.275(35)

$m_s = 97$ MeV/c²
95(09)

$m_\mu = 105.6583753$ MeV/c²
105.6583755(23)

$m_t = 173.0$ GeV/c²
173.0(04)

$m_\tau = 1776.9650$ MeV/c²
1776.86(12)

$m_b = 4.216$ GeV/c²
4.18(04)

This pattern matches our best measured values within 1 standard deviation in every case. Which, as far as I'm aware, lays

---

the nontrivial zero of the logarithmic integral $li(\mu) = 0$, or the vortex's throat), and an outer horizon boundary π. These equations are made exact with a small second order vortex effect, not included above for aesthetic simplicity.

$$3(m_N - m_P) = m_e (2^0)(\mu + 3 + \pi)\left(1 + \frac{Ж_1{}^2 Ж_5{}^2}{Ж_2{}^2 Ж_6{}^2}\right)$$

claim for the *best* example of matchup between theory and measurement in all of science![9]

They also give us the neutron/electron and proton/electron mass ratios, clearly broadcast Nature's critical condition for mixing 3 components of momentum between 2 boundaries $(\mu + 3 + \pi)$, and reveal the electron as the stable center of all mass projections.

$$\frac{m_N}{m_e} = \frac{1838.68366234}{1838.68366173(89)}$$

$$\frac{m_P}{m_e} = \frac{1836.15267335}{1836.15267343(11)}$$

The first 3 generations of this pattern relate to each other via vortex conditions—differing by multiples of 2 and breaking the symmetry of that arrangement once—with the 3^2. Likewise, the second triad of generations are related to each other via a power tower nesting of 2 and break the symmetry of that arrangement once with the $2m_s$.

When we compare this set of equations to matter's fundamental *charge* pattern, we immediately gain insight.

---

[9] These equations clearly align with the MS scheme's prediction for the value for the beauty quark mass (also called the bottom quark) over the 1S scheme. The extra precision in m_τ here comes from the family mass equations below. Neutrino masses are currently unknown, and therefore not included in this list.

$$(H \pm Z) = \pm W \qquad (0 \pm 1) = \pm 1$$
$$(N - P) = e \qquad (0 - 1) = -1$$
$$(d - u) = e \qquad \left(\left(-\tfrac{1}{3}\right) - \left(+\tfrac{2}{3}\right)\right) = -1$$
$$(s - c) = \mu \qquad \left(\left(-\tfrac{1}{3}\right) - \left(+\tfrac{2}{3}\right)\right) = -1$$
$$(b - t) = \tau \qquad \left(\left(-\tfrac{1}{3}\right) - \left(+\tfrac{2}{3}\right)\right) = -1$$
$$(\nu_\tau - \nu_\mu) = \nu_e \qquad (0 - 0) = \pm 0$$
$$(\nu_\mu - \nu_\tau) = \nu_e \qquad (0 - 0) = \pm 0$$
$$(\nu_\mu - \nu_e) = \nu_\tau \qquad (0 - 0) = \pm 0$$
$$(\nu_e - \nu_\mu) = \nu_\tau \qquad (0 - 0) = \pm 0$$
$$(\nu_\tau - \nu_e) = \nu_\mu \qquad (0 - 0) = \pm 0$$
$$(\nu_e - \nu_\tau) = \nu_\mu \qquad (0 - 0) = \pm 0$$

Nature's charge pattern is unmistakably orthogonal to the generation mass pattern—as its relations have zero dependence on scaling. Nevertheless, they are obviously otherwise simply connected, for when the generation mass pattern is stripped of its scale parameters, it is almost a perfect copy of the charge pattern. Only two adjustments upset the match. The first, is a flip *in* the (c, s, μ) generation.

$$(m_c - m_s) = m_\mu \qquad \rightarrow \qquad (s - c) = \mu$$

And the second, is the rotation *of* the (t, τ, b) generation.

$$(m_t - m_\tau) = m_b \qquad \rightarrow \qquad (b - t) = \tau$$

In other words, the generational mass and charge patterns of the fundamental particles transform into each other with just a flip *in* one group (a 2-member action) and a twist *of* another (a 3-member action)—the actions that underly the formation of a vortex.

This flip and twist happens to be exactly what is needed to rotate 6 (spin ½, charge 0) neutrinos into 3 spin zero, charge zero particles (3H) and 3 spin one, charge zero particles (3Z). This folds the distal ends of the generation pattern and the charge pattern into each other, closing a fractal loop.

How does that folding happen? How do 6 neutrinos get twisted into 3 Higgs and 3 Z bosons? The answer is found in how matter is arranged by *family*—which convey the rules for scale mixing.

$$\frac{m_H + m_Z + m_W}{(\sqrt{m_H} + \sqrt{m_Z} + \sqrt{m_W})^2} = \left(\frac{3}{2}\right)^{\frac{1}{2}} \left(\alpha_F - 1\right)^{-e^{2\gamma}}$$

$$\frac{m_e + m_\mu + m_\tau}{(\sqrt{m_e} + \sqrt{m_\mu} + \sqrt{m_\tau})^2} = \left(\frac{2}{3}\right)^{1} \left(\alpha_F - 1\right)^{e^{\pi i} + 1}$$

$$\frac{m_e}{m_p}\left(\frac{Ж_2}{Ж_1}\right)^2 = \left(\frac{2}{3}\right)^{2} \left(\alpha_F - 1\right)^{e^{2\pi i} + 1}$$

$$\frac{m_u + m_d + m_s}{(\sqrt{m_u} + \sqrt{m_d} + \sqrt{m_s})^2} = \frac{\left(\frac{2}{3}\right)^{2} + \left(\frac{2}{3}\right)^{1}}{2} \left(\alpha_F - 1\right)^{e^{3\pi i} + 1}$$

$$\frac{m_c + m_b + m_t}{(\sqrt{m_c} + \sqrt{m_b} + \sqrt{m_t})^2} = \left(\frac{2}{3}\right)^{3} \left(\alpha_F - 1\right)^{e^{-2\pi i} + 1}$$

$$\frac{m_{\nu_e} + m_{\nu_\mu} + m_{\nu_\tau}}{(\sqrt{m_{\nu_e}} + \sqrt{m_{\nu_\mu}} + \sqrt{m_{\nu_\tau}})^2} = \left(\frac{3}{2}\right)^{\frac{1}{2}} \left(\alpha_F - 1\right)^{e^{\pi i}}$$

Where α_F is the Feigenbaum Alpha constant, γ is the Euler-Mascheroni constant (which denotes the limiting difference between the natural logarithm and the harmonic series), and $Ж_1$, $Ж_2$ are the first and second Real solutions to our simple vortex equation.[10]

---

[10] With the second order reflection term included the full equations are

$$\frac{m_e}{m_p}\left(\frac{Ж_2}{Ж_1}\right)^2 = \left(\frac{2}{3}\right)^2 (\alpha_F - 1)^{e^{2\pi i} + 1}\left(1 - \frac{1}{2}\frac{Ж_1{}^2 \, Ж_5{}^2}{Ж_2{}^2 \, Ж_6{}^2}\right)$$

Using the mass values listed previously, these equations are satisfied with the following precision.

$$m_X = \frac{\text{Predicted}}{\text{Measured}}$$

(m_H, m_Z, m_W) family $= \dfrac{0.336339}{0.336334}$

(m_e, m_μ, m_τ) family $= \dfrac{0.666666442}{0.666666439}$

(m_e, m_p) vortex core $= \dfrac{1.00388092}{1.00388073}$

(m_u, m_d, m_s) family $= \dfrac{0.555}{0.572}$

(m_c, m_b, m_t) family $= \dfrac{0.66925}{0.66925}$

These family equations directly measure the scale symmetry/asymmetry between group members. If all particles in a family had equal mass, if there was no way to distinguish them by scale, then this relation would yield $\frac{1}{3}$. By contrast, if one

––––––––––––––––––

Neutrinos trivially maintain 6 charge arrangements because they have zero charge—until they are phase rotated into each other.

Props to Yoshio Koide for coming so close on this one.

family member greatly eclipsed the others in scale, this scaling relation would approach $\frac{3}{3}$.

Matter is trapped exactly in the middle of these two limits (at $\frac{2}{3}$), because matter is that which successfully balances Nature's 2 minimum boundaries against its 3 maximum boundaries.

Note the power inversions on the top and bottom family relations, and how the entire cascading sequence of mass, from minimum scale to maximum scale, is an elegant folding from the 3D world of our experience to the double cover 2D world of spinors (from $e^{\pi i}$ to $-e^{2\gamma}$). Beyond patterning the masses of the particles, this sequence codes the full translation between the world of our experience and the doubled-halved Möbius world of inversion.

Also note how matter particles grouped by family are phase-shift combinatorial. And how the prominence of $ж_1$ and $ж_2$ in this pattern (the 2 Real minimum solutions of the simple vortex equation)—leave no question as to what vortex is controlling these relations.

By *generation,* mass is organized by powers of 2^2. By *charge*, mass maintains a constant relationship on all 7 scales. And by *family*, mass is a measure of scale boundary mixing— coded by a tendency for period-doubling and bifurcation with α_F—*the* fundamental fractal scaling coefficient. These three properties are twisted together by the ordered union of the Planck boundaries—the limiting condition of vortex stability in a quantum fluid.

$$\frac{\cup\,(q_P, m_P, T_P)}{\cup\,(t_P, l_P)} = \frac{7}{2^2\,(\alpha_F)^{1/2}}$$

Now that we have accurate blueprints for matter's three orthogonal properties, let's root the scale of this entire pattern by expressing the electron mass in natural units (Planck mass units).

$$\frac{m_e}{m_{Planck}} = \left(\frac{8(e^\pi + 1)(1 \times 10^{-7})}{(\mu + 3 + \pi)}\right)^4 \left(\frac{(e + 2 + 2\pi)}{11}\right)^{1/2}$$

This equation tells a clear geometric story. The *mass* of the electron (its total distortive presence in the fluid) has two contributions: a localized wave, and its back reflection. These two parts are separated by 8 orders of magnitude, which means that they take up the entire range of Nature's scale-hyperbolic-octonion degrees of freedom.

$8(e^\pi + 1)$ communicates a fluid carved up into 8 hierarchal scales by quantization. On each scale, a rotation and reflection are taken as fundamental signatures of balanced action. (1×10^{-7}) tells us that 7 of those scales have broken scale-symmetry (the 1st and the 8th scales are symmetric).

To successfully form a vortex, Nature must trap three elements of momentum between two boundaries $(\mu + 3 + \pi)$, the zero and horizon boundaries of the vortex. To characterize the mass of that vortex, we simply divide the fluid's geometric structure by that stability condition and include its reflection.

Note how the reflection operator switches to an expression of balancing 2 things between e and 2π, spread over 11 dimensions. Because matter is that which successfully maintains a transition in dimensional resolution.[11]

---

[11] Recall that the union of the Planck constants is made exact with a similar operator.

$$\cup (t_P, l_P, q_P, m_P, T_P) = e\left(\frac{(e - 2 + 2\pi)}{7}\right)^{1/2}$$

Chapter 10
quantum mechanics

$\mathcal{N}$ ow that we have produced the exact properties of the fundamental particles of matter, standard quantum mechanics fantastically takes over to tell the rest of the story—detailing how atoms combine into molecules, which organize into crystals, cells, and with general relativity, all the way up to planets and solar systems and galaxies.

By what black magic does the story that stretches from the atomic scales to the galactic scales simply get attached to our ontology for free? No magic. No trick. Let me paraphrase the greatest minds on the topic and show you.[12]

To see how quantum mechanics is derived from our ontology, first we derive the Schrödinger equation; a task that begins with the insight that a fluid metric populated with particles, which collectively contribute to the evolution of a density field and a velocity field, necessarily has waves. These waves are analogous to sound waves, as their propagation is a function of the compressibility of the medium.

Therefore, under the simplest most general assumptions we can make, it becomes natural to model the evolution of our ideal non-viscous fluid as an acoustic metric. Which, of course,

---

[12] The mathematics in this chapter, as well as the discussion, follows the work of Detlef Dürr, Sheldon Goldstein, & Nino Zanghí in *Quantum Physics Without Quantum Philosophy*. The intellectual origins of the presentation herein trace back to a paper by the same authors titled *Quantum Equilibrium and the Origin of Absolute Uncertainty*. Journal of Statistical Physics. 67, 843-907 (1992), building on earlier works by Louis de Broglie and others.

means that there is going to be a wave equation playing the role of intrinsic descriptor to any system that is internal to that fluid.

We can derive this wave equation from the fact that a non-viscous fluid obeys energy conservation, de Broglie relations, and linearity.[13] Energy conservation can be expressed through the Hamiltonian H, which is classically used to represent the total energy E of a particle in terms of the sum of its kinetic energy T, and its potential energy V.

$$E = T + V = H$$

If a particle is restricted to one-dimensional motion its energy can be described as:

$$E = \frac{p^2}{2m} + V(x, t) = H$$

with position x, time t, mass m, and momentum p.

In three dimensions the equation becomes:

$$E = \frac{\boldsymbol{p} \cdot \boldsymbol{p}}{2m} + V(\boldsymbol{r}, t) = H$$

where $\boldsymbol{r}$ is the position vector and $\boldsymbol{p}$ is the momentum vector.

When we extend this formalism to represent more than one particle, we have to keep track of the fact that potential energy is a function of all the relative positions of the particles (versus being a simple sum of the particle's separate potential energies). Under this consideration, the general equation becomes:

---

[13] L. de Broglie. (1924). *Recherches sur la théorie des quanta* (Researches on the quantum theory), Thesis (Paris); (1925). *Ann. Phys.* (Paris) **3**, 22.

$$E = \sum_{n=1}^{N} \frac{\boldsymbol{p_n} \cdot \boldsymbol{p_n}}{2m_n} + V(\boldsymbol{r_1}, \boldsymbol{r_2}, \ldots \boldsymbol{r_N}, t) = H$$

Next up, de Broglie relations are captured by the fact that, in any acoustic metric the energy of any phonon (a collective excitation in the medium's elastic arrangements) is proportional to the frequency v (or angular frequency $\omega = 2\pi v$) of its corresponding quantum wave packet. This leads to the expectation that

$$E = hv = \hbar k$$

In this set up, any wave can be associated with a particle such that, in one dimension, the momentum p of the particle is related to the wavelength λ by

$$p = \frac{\hbar}{\lambda} = \hbar k$$

In three dimensions the relation becomes

$$\boldsymbol{p} = \frac{h}{\lambda} = \hbar \boldsymbol{k}$$

where $\boldsymbol{k}$ is the wave vector (the magnitude of $\boldsymbol{k}$ relates to the wavelength).

Next we achieve generality in our derivation by noting that in an acoustic metric, waves can be constructed via the sinusoidal superposition of plane waves. In other words, the superposition principle is expected to apply to the waves we mean to characterize, and more generally to any acoustic metric.

With these attributes we can structure an equation that characterizes the state of our fluid over time, without abandoning ontological clarity as to what that wave function is expressing.

To do this we construct an equation that is capable of translating the possible Hamiltonian of a system's particles (their possible kinetic and potential energies constrained by the aforementioned characteristics of the system) into a function of the state of the system, which we represent with ψ.

To capture the consequences of an internally quantized construction we express the phase of a plane wave as a complex phase factor via:

$$\psi = Ae^{i(\mathbf{k} \cdot \mathbf{r} - \omega t)} = Ae^{i(\mathbf{p} \cdot \mathbf{r} - Et)/\hbar}$$

Next we take the first order partial derivatives of this equation with respect to space and time:

$$\nabla \psi = \frac{i}{\hbar} \mathbf{p} Ae^{i(\mathbf{p} \cdot \mathbf{r} - Et)/\hbar} = \frac{i}{\hbar} \mathbf{p} \psi$$

$$\frac{\partial \psi}{\partial t} = -\frac{iE}{\hbar} Ae^{i(\mathbf{p} \cdot \mathbf{r} - Et)/\hbar} = -\frac{iE}{\hbar} \psi$$

When we combine this with our earlier expression for energy, this leads to:

$$-i\hbar \nabla \psi = \mathbf{p} \psi \longrightarrow \frac{\hbar^2}{2m} \nabla^2 \psi = \frac{1}{2m} \mathbf{p} \cdot \mathbf{p} \psi$$

$$\frac{\partial \psi}{\partial t} = -\frac{iE}{\hbar} \psi \longrightarrow i\hbar \frac{\partial \psi}{\partial t} = E\psi$$

Now if we multiply our three-dimensional energy equation by ψ we get:

$$E = \frac{\boldsymbol{p} \cdot \boldsymbol{p}}{2m} + V \longrightarrow E\psi = \frac{\boldsymbol{p} \cdot \boldsymbol{p}}{2m}\psi + V\psi$$

which naturally collapses into the following wave equation:

$$i\hbar \frac{\partial \psi}{\partial t} = -\frac{\hbar^2}{2m}\nabla^2\psi + V\psi$$

This equation is exactly what Schrödinger came up with. Nevertheless, as it stands, it is slightly incomplete. To capture the full range of possibilities present in the actual fluid situation, this equation must be augmented to allow for inter-particle interactions.

To get there, we characterize a Bose-Einstein Condensate (BEC) whose state can be described by the wave function of the condensate ψ. The particle density of this system would then be represented by $|\psi|^2$.

If all of the particles in this fluid have condensed to the ground state, then the total number of atoms N in the system will be $N = \int |\psi|^2 \, d\vec{r}$. Treating these particles as bosons, and using mean field theory, the energy E associated with the state ψ takes on the following expression:

$$E = \int d\vec{r} \left[\frac{\hbar^2}{2m}|\nabla\psi|^2 + V|\psi|^2 + \frac{1}{2}U_0|\psi|^4 \right]$$

By minimizing this expression with respect to infinitesimal variations in ψ, and fixing the number of particles in the system, we end up with the Gross-Pitaevskii equation (also called the nonlinear Schrödinger equation, or the time-dependent Landau-Ginzburg equation).

$$i\hbar \frac{\partial \psi}{\partial t} = \left(-\frac{\hbar^2 \nabla^2}{2m} + V + U_0 |\psi|^2 \right) \psi$$

where: m is the mass of the bosons, ψ is the external potential, and U_0 represents inter-particle interactions.

Note that when inter-particle interactions go to zero this equation reduces to Schrödinger's original equation.

Deriving the wave equation from first principles like this offers us an unprecedented ability to ontologically access what the wave equation means and represents. The wave equation characterizes the possible state evolutions that the substrate of reality can support, including what internal periodic excitations (particles) are allowed to stabilize, and the interactive properties of those emergent stabilities. In short, the wave equation describes the flexible character of the fluid medium of reality.

Now that we've got the Schrödinger equation as the dynamic descriptor of our system, quantum mechanics can be completed *in full* by a mere specification of actual particle positions, which evolve (in configuration space) according to the *guiding equation*.

A full description of a system composed of N particles naturally includes a specification of both the particle positions and the wave function (Q, ψ), where

$$Q = (Q_1, Q_2, Q_3, \dots Q_N) \quad \in \mathbb{R}^{3N}$$

is the configuration of the system and

$$\psi = \psi(q) = \psi(q_1, q_2, q_3, \dots q_N)$$

a (normalized) function on configuration space—is its wave

function.

With this complete state specification in hand (Q, ψ) all that is left to do is specify a law of motion. The simplest choice we can make here is one that is causally connected. In other words, one whose future is determined by its present specification, and more specifically, whose average total state remains fixed.

Of course, to choreograph this stage we call in our wave equation (our evolution equation—the Schrödinger equation),

$$i\hbar \frac{\partial \psi}{\partial t} = H\psi_t = -\frac{\hbar^2}{2m_k} \nabla^2 q_k \psi_t + V\psi_t$$

In keeping with our previous considerations, the evolution equation for Q should be $\frac{dQ_t}{dt} = v^{\psi t}$ with $v^\psi = \left(v_1^\psi, v_2^\psi, v_3^\psi, \dots v_N^\psi\right)$ where v^ψ takes the form of a (velocity) vector field on our chosen configuration space $\mathbb{R}^{3N}$.

Under this set up the wave function ψ clearly reflects the motions of the particles in our system, in a macroscopic averaged-over sense, based on the underlying assumption of elastic interaction, or zero viscosity. These motions are coordinated through a vector field that is defined on our specified configuration space.

$$\psi \longrightarrow v^\psi$$

If we are requiring time-reverse symmetry and simplicity to hold in our system then,

$$v_k^\psi = \frac{\hbar}{m_k} Im \frac{\nabla q_k \psi}{\psi}$$

Notice that there are no ambiguities here. The gradient ∇ on the right-hand side is suggested by rotation invariance, the ψ

in the denominator is a consequence of homogeneity (a direct result of the fact that the wave function is to be understood projectively, which is in turn an understanding required for the Galilean invariance of Schrödinger's equation alone), the *Im* by time-reverse symmetry, which is implemented on ψ by complex conjugation in keeping with Schrödinger's equation, and the constant in front falls directly out of the requirements for covariance under Galilean boosts.[14]

Therefore, the evolution equation for Q is

$$\frac{dQ_k}{dt} = v_k^\psi(Q_1, Q_2, Q_3, \ldots Q_N) \equiv Im\frac{\nabla q_k\psi}{\psi}(Q_1, Q_2, Q_3, \ldots Q_N)$$

This completes the pilot-wave theory that Louis de Broglie and David Bohm constructed in 1952.[15] As it stands, it exhaustively depicts a nonrelativistic universe of N particles without spin.[16] To account for all the apparently paradoxical quantum phenomena associated with spin, we just retain the complex conjugate of the wave function.

For considerations without spin the complex conjugate of the wave function cancels because it appears in the numerator and the denominator of the equation. The full form of the evolution equation is

---

[14] Detlef Dürr, Sheldon Goldstein, & Nino Zanghí. *Quantum Physics Without Quantum Philosophy*, pp. 5–6.

[15] David Bohm. (1952).

[16] Of course, in the limit $\frac{\hbar}{m} \to 0$, the Bohm motion Q_t approaches the classical motion. See: D. Bohm & B. Hiley. (1993). *The Undivided Universe: an Ontological Interpretation of Quantum Theory.* Routledge & Kegan Paul, London; Detlef Dürr, Sheldon Goldstein, & Nino Zanghí. *Quantum Physics Without Quantum Philosophy*, p. 7.

$$\frac{dQ_k}{dt} = \frac{\hbar}{m_k} Im \left[\frac{\psi^* \partial_k \psi}{\psi^* \psi} \right] (Q_1, Q_2, Q_3, \dots Q_N)$$

Notice that the right-hand side of the guiding equation is the ratio for the quantum probability current to the quantum probability density.[17]

Given that the classical formula for current is density times velocity, "it requires no imagination whatsoever to guess the guiding equation from Schrödinger's equation."[18]

This simplicity speaks to why the de Broglie-Bohm theory has been said to be "the most naïvely obvious embedding imaginable of Schrödinger's equation into a completely coherent physical theory."[19] It elegantly restores determinism into the dynamics of physical reality; accounting for all the phenomena governed by non-relativistic quantum mechanics—"from spectral lines and scattering theory to superconductivity, the quantum hall effect, quantum tunneling, nonlocality, and quantum computing."[20] And it does so under assumptions of greatest symmetry.

---

[17] Sheldon Goldstein. *Bohmian Mechanics*. For further examples of how easily spin can be dealt with in the Bohmian formalism see: J. S. Bell, 1966, 447–452; D. Bohm, 1952, 166–193; D. Dürr et all, A survey of Bohmian mechanics, Il Nuovo Vimento, and Bohmian mechanics, identical particles, parastatistics, and anyons, In preparation.
[18] https://plato.stanford.edu/entries/qm-bohm/
[19] D. Dürr, et al.
[20] https://plato.stanford.edu/entries/qm-bohm/

Chapter 11
twist and reflection

"The mathematician's patterns, like the painter's or the poet's, must be beautiful; the ideas, like the colors or the words, must fit together in a harmonious way. Beauty is the first test: there is no permanent place in this world for ugly mathematics."
  G. H. Hardy

On the quantum scale there are two unique geometric actions—twists and reflections. And twisting half-way around (by pi radians) $\left(e^{\pi i}\right)$ is exactly equal to undergoing full reflection (-1).

$$e^{\pi i} = -1$$

To see the full consequences of this equality, let's rearrange this equation so that it reads, "Two different geometric actions combine deconstructivity to perfectly cancel each other out."

$$e^{\pi i} + 1 = 0$$

Generalizing this *deconstructive* condition

$$e^{x\pi i} + 1 = 0$$

reveals that it holds equally for $x = -1$, $+1$, and 3.

$$e^{-1\pi i} + 1 = 0 \qquad e^{+1\pi i} + 1 = 0$$

$$e^{3\pi i} + 1 = 0$$

Conversely, constructive combinations

$$e^{x\pi i} + 1 = 2$$

hold for $x = -2$, 0, and $+2$.

$$e^{-2\pi i} + 1 = 2 \qquad e^{0\pi i} + 1 = 2$$

$$e^{+2\pi i} + 1 = 2$$

When these constructive combinations are achieved, their boundary conditions carve up the world in a way that treats 0's and 2's as interchangeable.[21] This captures the full condition of quantum action for matter.

$$e^{x\pi i} + 1 = \begin{cases} 0 \\ 2 \end{cases}$$

These constructive and deconstructive conditions divide the world up into ordered composites of $(-2, 0, +2)$ and $(-1, +1, 3)$—sets whose ordered elements differ simply by steps of 2.

Under quaternion exponentiation, where rotations involve three orthogonal directions, this identity becomes

$$e^{\pi(i \pm j \pm k)/\sqrt{3}} + 1 = \begin{cases} 0 \\ 2 \end{cases}$$

---

[21] See the Riemann Hypothesis.

where $\{i, j, k\}$ are the basis elements. The full octonion identity is

$$e^{\pi(a_1 i_1 + a_2 i_2 + a_3 i_3 + a_4 i_4 + a_5 i_5 + a_6 i_6 + a_7 i_7)} + 1 = \begin{cases} 0 \\ 2 \end{cases}$$

with basis elements $\{i_1, i_2, i_3, i_4, i_5, i_6, i_7\}$, and real coefficients whose squares sum to 1.

$$a_1{}^2 + a_2{}^2 + a_3{}^2 + a_4{}^2 + a_5{}^2 + a_6{}^2 + a_7{}^2 = 1$$

Applying the structural condition of possessing 7 breaks of scale symmetry,

$$a_1{}^2 < a_2{}^2 < a_3{}^2 < a_4{}^2 < a_5{}^2 < a_6{}^2 < a_7{}^2$$

casts our substrate as *scale*-hyperbolic-octonion.

The octonions are widely considered the most elegant mathematical structure possible. Not only are they the largest normed division algebra, they are also the only fully closed division algebra. For these reasons, many luminaries have long suspected that one day the "octonions will ultimately be seen as the key to a unified field theory in physics."[22] That day is today.

In a braid of complex reflection, Nature unveils the ultimate twist—its perfection. We now know what is really needed, the bare minimum of form required to make a universe whose patterns give rise to emergent qualities; imbuing reality with kinds of interactions—kinds of dynamics—that impregnate

---

[22] Tevian Dray & Corinne Manogus, *The Geometry of the Octonions*. For those eager for more math here, I recommend reading Hyperbolic Octonion Formulation of the Fluid Maxwell Equations, by Süleyman Demir and Murat Tansli. Journal of the Korean Physical Society, Vol. 68, No. 5, March 2016, pp. 616~623.

existence with all the fundamental tendencies a universe needs to end up with, to produce the ordered complexity we see around us.

Instead of calling everything that modern physics has yet to explain into existence (photons—electromagnetic fields; the fundamental particles of matter and their quantized values of spin, charge and mass—and their gravitational fields; the interactive tendencies of those particles of matter—the strong and weak nuclear fields; all of the constants of Nature; and dark matter; and dark energy), instead of giving all of those things primary ontological status, we automatically get all of those properties, literally all of them, from the cascade of broken symmetry defined by the boundaries of a quantum fluid.

Constructing a universe this way represents a *least action kind of principle* on the *ontological* level—achieving the minimization at which Occam's razor aims. The best solution is the simplest foundation conceivably possible that gives rise to the nouns and verbs we need to explain—all on its own. For the first time in history, we now have a solution that counts as an answer—one that gives us visceral access to all the forms of reality and their geometric origins.

According to that solution, we are in *a perfect universe*: where the world is made beautiful by paying attention, and the magnitude of our growth depends on the beauty of our vision.

Acknowledgements

I thank George Cassiday for the passion of his ETI class.

I thank Albert Einstein for teaching me true courage of aim, Leonhard Euler for his insightful reach, and Gene Shoemaker (the man on the moon) for his bold example.

I thank my Num for pinky holds, Shangri-La, and for really meaning it.

I thank my $\sqrt{-1}$ Cloud for her delightful imagination.

I thank Elaine, Phil, and Matt Emmi for the way they gave me a key.

I thank David Heggli for being my Huxley—for tirelessly reaching to understand more, making this quest a priority, and for believing that maintaining an open-minded discussion, for however long it takes, is what friends do.

I thank Jeff Chapple—for all the Alfred stuff, and David Cantu for what may or may not have been said in the cone of silence.

I thank Michael Faraday, Benoit Mandelbrot, Louis de Broglie, David Bohm, Stephen Wolfram, Shelly Goldstein, Detlef Dürr, Nino Zanghí, Grant Sanderson (3Blue1Brown), Derek Muller (Veritasium), Brady Haran (Numberphile), Franck Laloë, Garrett Lisi (PSI), Burkard Polster (Mathologer), Destin Sandlin (Smarter Every Day), Dianna Cowern (Physics Girl), Grigori

Volovik, Robert Brady, Ross Anderson, Erwin Madelung, Tim Maudlin, Richard Feynman, Craig Callendar, Christian Wüthrich, Cohl Furey, Yoshio Koide, David Richeson, William Thomson, TED, and The Royal Institution—for their gifts of clarity.

And I thank William Prince, John Gilbert, Wayne Eskridge, Richard Pape, Marcus Tofanelli, Tanja Traxler, Avi Rubin, and Pierre Rousseau for their editorial feedback.

Finally, I thank Hans de Vries for his observation, and Philipp Preetz for pointing me to it.

To support this kind of independent scientific research, please find me at www.vortex.institute.

Appendix Ж7

$$\frac{1}{\text{Ж}} + \text{Ж} + \frac{\text{Ж}^3}{2\pi} = \left(i^i\right)^{-\pi/2} - m_P$$

The simple vortex equation derived in Chapter 7 has 8 limiting solutions. On the maximum scale, where $m_P = 2.176435 \times 10^{-8}$, it possesses 2 real solutions $(\text{Ж}_1, \text{Ж}_2)$, and 2 complex solutions $(\text{Ж}_3, \text{Ж}_4)$. Likewise, the minimum scale (where $m_P = 1$) has 2 real solutions $(\text{Ж}_5, \text{Ж}_6)$ and 2 complex solutions $(\text{Ж}_7, \text{Ж}_8)$. They are:

$\text{Ж}_1 = 0.0854245431533310(18)$
$\text{Ж}_2 = 3.667567534854999(33)$
$\text{Ж}_3 = -1.876496039004165(16) + 4.06615262615971(03)\,i$
$\text{Ж}_4 = -1.876496039004165(16) - 4.06615262615971(03)\,i$

$\text{Ж}_5 = 0.093474041000289951749648\ldots$
$\text{Ж}_6 = 3.526755530609707752174473\ldots$
$\text{Ж}_7 = -1.810114785804998851 - 3.97279144047142328\,i$
$\text{Ж}_8 = -1.810114785804998851 + 3.97279144047142322\,i$

These solutions possess the following combinatorial properties.

$$\text{Ж}_1 + \text{Ж}_2 + \text{Ж}_3 + \text{Ж}_4 = 0$$
$$\text{Ж}_5 + \text{Ж}_6 + \text{Ж}_7 + \text{Ж}_8 = 0$$

$$\text{Ж}_1{}^2 + \text{Ж}_2{}^2 + \text{Ж}_3{}^2 + \text{Ж}_4{}^2 = -4\pi$$
$$\text{Ж}_5{}^2 + \text{Ж}_6{}^2 + \text{Ж}_7{}^2 + \text{Ж}_8{}^2 = -4\pi$$

$$\text{Ж}_1{}^4 + \text{Ж}_2{}^4 + \text{Ж}_3{}^4 + \text{Ж}_4{}^4 = -8\pi + 48\,\zeta(2)$$
$$\text{Ж}_5{}^4 + \text{Ж}_6{}^4 + \text{Ж}_7{}^4 + \text{Ж}_8{}^4 = -8\pi + 48\,\zeta(2)$$

$$\sqrt{\text{Ж}_2{}^2 - \text{Ж}_1{}^2} = 2^8 - 48\,V_h^*$$

$$\sqrt{\text{Ж}_2{}^2 + \text{Ж}_1{}^2} = 8\,e - \frac{(2^8 - 1)\,W(1)}{8}$$

$$\frac{\text{Ж}_2\text{Ж}_5}{\text{Ж}_1\text{Ж}_6} = \frac{11V_h^* + 48}{(2^2 + 5^2 + 8^2)}$$

Where $\zeta(s)$ is the Riemann zeta function, V_h^* is the dimension in which an n-hypersphere has maximal volume, and $W(1)$ is the Omega one constant, which is the unique real constant that satisfies the equation $\Omega e^{\Omega} = 1$.

While writing this book I filled 3 notebooks with additional interesting and potentially insightful equations. If you have absorbed the contents of this book (and Einstein's Intuition) and are interested in more information, or if you would like to help us further animate these concepts, please join the discussion at www.vortex.institute.

Other books by Thad:
- Einstein's Intuition:
    Visualizing Nature in Eleven Dimensions
- Moon Rock: Mare Crisium
- Passages

www.ingramcontent.com/pod-product-compliance
Lightning Source LLC
Chambersburg PA
CBHW041711200326
41518CB00001B/154